CW01431757

CLIMATE CRISIS

CLIMATE CRISIS

Finding Hope in a Changing World

AURORA WINTERS

QuillQuest Publishers

CONTENTS

CHAPTER 1

Introduction

You may say our time here has not been golden. Years later, teen-agers of today might be reliving their year of alienation from friends and adventure during the 2020 pandemic fight and wonder how the world began to heal, to restore. Our actions have the power to bring some of the Earth's lost biological beauty back. Our activities can make the world glow green. And young conservationists of today, like Fezile Gushu, 12, are ready for the challenge. Have you ever woken up early in the morning to that soothing silence of nature? I hardly ever did but I long for it now. It's amazing to see nature just getting on with it, the lions are more active, there are little bushes everywhere, and swarms of swallows are catching insects at sunrise. The Earth is a powerful force, an ecosystem we should all be willing to protect.

The troubling impact of the climate crisis can lead to feelings of despair. For Lizzie Daly, feelings of hopelessness were quelled when she encountered a small array of hope right on her doorstep in the midst of a pandemic. Filled with inspiration to alleviate the pain inflicted on our Earth, she embarked on a mission to collate stories from across the globe – stories of hope and recovery, of an unexpected positivity, curiosity, and activity among wildlife as a result of the pandemic. And stories of hope from future conservationists, the young people whose movement has been delayed but not derailed.

CHAPTER 2

Understanding the Climate Crisis

Ultimately, our attention must shift to the impacts of climate change that are harming every corner of life on earth. It is the consequences of climate change that we must address while avoiding the drivers of this change.

Widespread heatwaves affect the young, elderly, and those who are unable to escape the excessively hot conditions. The climate crisis also can impact personal finances. Insurance rates increase in conjunction with increased damage from floods, hail, wind, and fire. Medical costs are increased by asthma, cardiovascular stress, vector-borne diseases, waterborne diseases, and injuries from storms with side effects. Evacuations are costly, damage to your property even more so. You may pay more for food as the price of crops increases because of shortened growing seasons with severe damage to crops caused by a variety of weather and climate conditions.

The weather is changing, becoming more violent and unpredictable. Hurricanes are more powerful. Tornadoes are wider, more destructive, and found in places where they are rarely seen. Floods and prolonged periods of drought are both increasing, both with severe impacts on communities. Meanwhile, temperatures are hotter or colder, with the extreme temperature variation causing some farmers to wonder when they should plant their summer crops. More widespread heavy precipitation can hurt farmers, golfers, and field workers. Crops can be destroyed through drought, heat, and floods.

The climate crisis portends worldwide issues that may directly impact you and your family. Many of these can be resolved and lessen the momentum of the crisis or mitigate the pain. Your understanding, with our help, can make a world of difference.

CHAPTER 3

The Impact of Climate Change

However, there is so much positive that more and more businesses are investing in and discovering than in all types of other industries, that working to prevent climate damage (often through GHG reduction, energy efficiency, renewable energy, supporting policy changes at local, state and federal levels) is often a fantastic, triple-bottom-line (people, planet, profit) business decision. I will also add that progress in social fairness and poverty alleviation are also a major component of helping the world achieve climate stability. Climate and our interconnected world almost always works in this way. If done right, our efforts can make the world a better place in so many complementary ways. To me, that is the sign of a living, interconnected world that is calling us to our higher selves. The concerns people feel, the debating we do, even the arguing about the reality of the situation are all about value - about whose values are primary, as well as whose are right. What organizations publicly 'value' sometimes means far more than how they behave.

Clear signals turn conversations about common values into real, practical action.

The impact of climate change has more aspects and nuances than we often consider. Making sure the impact is not overlooked has been my primary work over the past 15 years, speaking at hundreds of conferences for companies, non-profits, schools, energy utility, consulting firms, all types of organizations around the world on the topic of the business impact of climate change. The amazing thing is that very few organizations in most industries today fully understand the financial impact on the business unless you are in gas, oil, coal, agriculture or insurance - standalone, huge industries. As important as the physical effects are, they are not being considered by the business world. I do not mean the business perception of the impact. I do not mean business sensation of the impact. I mean the unemotional, practical, ROI based, doing the responsible thing - understanding the impact. Truly a tragedy.

CHAPTER 4

Global Efforts in Combating Climate Change

The Kyoto Protocol, which was signed in 1997 and implemented some years later, was one of the methods through which Convention aims were to be executed. The Protocol bound major emitters to reduce the rates of greenhouse gases that induce global warming by 5% in the commitment period of 2008-2012 compared to the emission rates that were in place in 1990. While the mechanism of the market and other devices allowed further emission reduction in a cost-effective way. In 2012, a new mechanism called the Doha Amendment was then agreed upon to set down the principle of the second commitment period. At additional stages, the United Nations Framework Convention on Climate Change had also initiated the growth of its operating procedures, including the supporting research within the field of climate change and for international financial resources for both adaptation and transition to a low-carbon economy.

7

Climate change represents the greatest environmental, social, and economic crisis in human history. Tackling such phenomenal issues requires collaboration, action, and policy responses at global, regional, and local levels. Collective cooperation has played a role in designing a specific set of international measures, ultimately reaching a global agreement in 1992 when the Framework Convention on Climate Change was signed at the United Nations Earth Summit in Rio de Janeiro. The Foundational Convention has been ratified by nearly 200 member countries. The Convention calls for the stabilization of dangerous levels of emission of greenhouse gases that cause change in our atmosphere. To achieve this, all countries, whether the largest or smallest, and regardless of their development status, should take action to reduce their greenhouse gas emissions.

CHAPTER 5

Individual Actions for a Sustainable Future

We don't just need a plan, but an all-encompassing plan that is best suited for the person, the situation, and societal trends that could disrupt the person's plans and progress. There isn't a "one-size-fits-all" approach to promoting sustainable behaviors. Marker moderates the relation between a social norm on PF and moral norm on PF, IA, and social norms on IA. People are less likely to enact sustainable behaviors if they take climate or moral issues less seriously compared to their neighbors. If they are organized, educated, and live near environmentally friendly options, people are more likely to live sustainably. Moral NF had a weaker correlation with IA when markers' value on either IA or PF were low. This can occur if the association dampens as the two variables approach zero, or as the scatterplot does imply, the measure of means value on both IA and PF.

This chapter discusses the urgent and necessary transition to sustainability, with individual actions aimed at forging a path towards the sustainable behaviors needed to tackle the problem.

In a 2019 poll, roughly half of us agreed that climate change was a crisis, with 69 percent of young people agreeing. However, it's difficult to know the "right thing" for us as individuals to do. Behaviors that we are told we should engage in, like eating less meat or driving more fuel-efficient cars, can feel like doing our part for the environment while we wait for others to do something too. However, individual change has the potential to mobilize collective action, as it leverages social influence and the infrastructure of networks.

Climate change remains the most significant challenge humans have had to face, with our own survival hanging in the balance. According to a March 2020 poll by the Yale Program on Climate Communication, 73 percent of Americans believe that global warming is happening and 62 percent are alarmed or concerned. However, despite the message that "every action counts," we do little other than talk about the weather.

CHAPTER 6

Innovations and Technologies for Climate Solutions

Mr. Jabr has given us hope in that he is not too worried about technology that may not have been thought of yet that could help us in our race to transform the economy and reduce or eliminate greenhouse gases. "The rapid scaling of technologies that already exist in some form or another is the most urgent priority". Not sure where this ranking of priorities is coming from because the intersection of the new and unknown are vital for our upward trend responses to cosmic pressure. Experiments with concentrating sunlight to produce steam have shown potentials all around and it's attractive because its work doesn't have to be stored in a battery or emitted from a power cell. In 2015, the Parker Solar Probe was launched; it's probing the full fury of the Sun's radiation only 3.9 million miles from its surface where temperatures are three times higher than in a typical Star Wars treatment. The heat that is absorbed by the 4.5 inch carbon shield that protects the probe is so intense, the thermal imbalances shift at rates of tens of thousands of degrees Fahrenheit over time intervals of only a few hours. This presents situations that the scale-to-power technology activists can't ever predict.

Modern technology is an amazing thing, and every day new inventions are made which will make our lives easier and better. People are not as familiar with the incredible inventions that have been developed as solutions to our world's climate crisis as they are with the latest iPhone version or other gadgets. But climate solutions do exist, and Ferris Jabr has done a good job describing some of these in an April 2021 article in Effect. He has also pointed out that some people don't share information about these innovations because they're afraid it could squelch people's motivation to act and changes may actually become even more urgent in the future. "Tolerance of climate anxiety has become a mark of both moral virtue and intellectual seriousness - a kind of courage. Any sense of hopefulness, by contrast, suggests naivety, detached from reality". No time to be naive or detached.

CHAPTER 7

Adapting to a Changing World

The problem with solving the problem early, though, is we often don't recognize the problem before a number of long-term symptoms stemming from the root cause have begun to ruin our lives. The most wicked symptoms of the climate crisis aren't expected in earnest for many decades; perhaps none until after this author is dead and gone. Without symptoms, the climate crisis might be even more difficult to surmount. Short-term crises may be easier to handle, whereas long-term crises can erode our hopes and even our very spirits. But the thought experiments of our greatest existential scientists can help us design and implement a politically unifying response.

How climatologists think we ought to solve a problem like a melting ice sheet. It's commonly said the best time to fix a problem is before it becomes a crisis. The story assumes that the worst time to address the problem is after it begins. This is an obvious lesson to apply to the climate crisis. To save the planet, the problem should be addressed long before the problem becomes viral. There is, after all, a point of no return for any effort to head off any problem. After that point is reached, the problem cannot be stopped, except perhaps by taking radical action against a symptom of the problem. Against the climate crisis, for instance, reducing the sunlight reaching earth can cool the planet, even if it exacerbates the problem. Looking at the problem more constructively, there's also a point of diminishing returns for any solution. Beyond that point, the solution usually becomes a cause of some other problem, and thus it can no longer be considered as a net solution. Addressing such problems early would, therefore, be more effective and less destructive.

CHAPTER 8

The Role of Government and Policy Making

For the most part, market forces will drive the use of alternative sources of power, including wind, tidal, and solar, with some government regulation needed for incentives and possibly also mandates. When power supplies are plentiful and widespread, the cost can be reduced so that it will also be cheaper to use electricity in preference to diesel and petrol. There is also a need for a replacement for aviation fuel. The United Kingdom Government made climate change policy a priority by being the first major economy in the world to set a legally binding target of a reduction of 80 percent in greenhouse gas emissions from 1990 levels by 2050. A Climate Change Act (2008) was brought forward, which addressed a range of policy measures across different areas and encouraged the growth in renewable power whilst aiming for energy efficiency. This target has continued to be supported since. Unfortunately, the government's Climate Change Committee has warned that instead of reducing emissions, the United Kingdom has instead been taking credits from carbon credits from other countries, thereby disguising the increase in emissions. The worry is that instead of meeting targets purely through UK-based greenhouse gas reduction, the UK has been treating the carbon credits of other nations as its own with no knock-on effect on its own emissions.

Government is needed to address the current climate crisis with the power and authority that it can bring to the largest problems faced by humanity. Each of us acting individually and locally can help, but we need government to drive much of the change that is needed. The largest contributor to greenhouse gases comes from the energy and transport systems, and government policies can have a significant impact on both. There are presently few economic barriers to widespread energy conservation and development of renewable power; the inaction and resistance of vested interests has been a stumbling block.

CHAPTER 9

Education and Awareness on Climate Change

According to a recent global study, the percentage of people who believe that climate change is real has increased in certain countries but is much lower than the countries examined. Results showed that Japan had one of the lowest levels of climate beliefs compared to other countries. The Re-Energizing the Public Health-Free Strategy and Intensive Climate Social Sciences analysis advocate for initiatives, strategies, and organizational innovations aimed at both the general public and the media. It is widely recognized that treating climate change with passive strategies is not as effective as treating it with scientifically proven strategies. The hope is to continue to gather information and suggestions from social scientists who are interested in climate change in order to advance the field further, make scientifically relevant and appropriate choices, and promote public understanding of climate change.

Members of the scientific and environmentalist community have been writing and speaking about ways to combat the climate crisis for over half a century. Earth Day, which was first celebrated on

April 22, 1970, in the United States, took place in hundreds of communities and brought millions of participants to share their desire to protect the environment. Many local, national, and global measures dealing with emissions, reductions, and alternative fuels have taken place as a result of the first Earth Day, and although progress has been made, the climate crisis is an ongoing societal issue threatening the future of humanity.

CHAPTER 10

Sustainable Agriculture and Food Systems

The good news is that it is entirely feasible to achieve this 30% reduction in greenhouse gas emissions through small, medium, and large-scale farmers in all nations (with the exception of the richest 20% in terms of GHG emissions, including China and India), while also enhancing ecosystems and biodiversity. This can be done right away - TODAY - using the sustainable agriculture practices currently available at our joke. According to the Rodale Institute Farm Systems Trial, the world could reduce emissions to about 45 GtCO2 equivalents simply by changing agricultural practices on around 37% of world cropland. Bearing in mind that 80% of the world's food is produced by a family farm, it is fortunate indeed that the sustainable, organic practices we need are already in the hands of the world's most abundant farmers - hundreds of millions of smallholders.

Sustainable, resilient farming and food systems can play a major role in helping the climate crisis because these activities can simultaneously: - Sequester carbon in soils - Reduce greenhouse gas emissions - Prevent deforestation and habitat destruction - Increase local people's self-sufficiency and enhance communities and economies.

Climate change and its impacts are decimating animal and plant populations, as well as human communities and cultures worldwide. Yet our current food production system is responsible for almost 30% of global greenhouse gas emissions - the same share as the entire transportation sector! At the same time, it is also responsible for large negative impacts on ecosystems and biodiversity, depletion of natural resources, and social and economic inequity. Needless to say, our food systems are not currently fit for purpose for the planet nor its people.

CHAPTER 11

Preserving Biodiversity and Ecosystems

Earth's biological interactions are central to its ability to support life, regulate atmospheric composition, and maintain habitable environments. Considerable disservice has been done by portraying ecosystems as scientists often do: as networks. Networks emphasize interactions between pairs of species, as if species exert direct positive or negative influences on one another. Actually, indirect interactions often exceed direct ones. These occur because some species, called 'keystone' species, or because particular trophic architectures, called 'trophic pyramids,' concentrate and amplify or reduce negative interactions between other species - in particular competitive or predator-prey interactions. In experiments and mathematical models, predictability and stability of ecosystems was found to increase with these aspects of ecosystem complexity. Further, as ecosystems grow more complex by incorporating many species and evolutionary innovations, they display an enhanced ability to recover from environmental disturbances.

Over the last century, human activities have affected animal and plant life more profoundly than during any other period in human history. One study estimated that up to 1 million species of animals and plants are threatened with extinction - many of which are expected to disappear in the next several decades. Today, the most highly endangered or charismatic of the world's fauna receive media attention and support as the focus of conservation efforts. This is understandable but incomplete. Our planet would not be able to support us if ecological relationships that sustain us were disconnected, nor if ecosystems were replaced by monocultures or unproductive wastelands.

CHAPTER 12

Renewable Energy Sources and Transition

After we adopt renewable energy sources, what kind of world can we look forward to? Will our lives be diminished by the transition to electricity, heat, and motive force provided by the wind, sun, rain, tides, and the heat of the Earth? Or might human ingenuity liberating renewable energy sources from their insidious chains usher in an era when the incomes of poor countries and peoples rise by an order of magnitude, when the wealth and employment potential of the larger portions of the human family are finally seen as not less, but more than that of the present drop or two of humanity that supports itself by the overuse of unsustainable sources? When the transition to renewable resources liberates the human family from a military-economic system worthy of the worst of the Inquisition popes, when investments returning significantly less than the cost of fossil fuels can and will no longer be made, I have no doubt that the problem of militarism and old media will be behind us, being a question of money and power.

By their very nature, renewable sources of energy are widely dispersed across the Earth, making them practical sources of energy for long-term human habitation of Earth. Solar energy comes not just from our nearest star, but from nuclear reactions in stars many light years distant, giving society an effectively infinite supply of energy. Thoughtful placement of renewable energy facilities can allow the natural ecosystem to continue to flourish while retrieving these practical sources of electricity, heat, and mechanical work to enable healthy human cultures. Even with a little human effort, life finds a way, all by itself. Crimea, Area beta, were difficult environments for life to return to after devastating fires. Yet, with recent rains, the land has already greened.

CHAPTER 13

Sustainable Transportation and Infrastructure

Infrastructure such as roads, sewers, water systems, airports, streetlights and public facilities, etc. must be built more sustainably in the future, reflecting the use of sustainable materials and accounting for user costs over the life-cycle of such infrastructure, including the emissions and resource use required to construct as well as the emissions from the activities which the infrastructure supports. The sustainability of infrastructure is rapidly becoming an issue, as reflected in the publication of a high-profile report on the subject by the Canadian Council of Chief Executives (CCCE).

This represents a critical sector for a number of reasons, including not only the need to reduce emissions, but also the natural synergy between taking action to reduce emissions and actions that will help us to deal with the looming peak in world oil production.

The transportation and building sectors in Canada combined are now responsible for about 52% of Canada's GHG emissions. We need to transform our transportation sector and our buildings to cut emissions and to ensure that our transportation system is prepared to address oil scarcity. One of the best ways to do this is through smart growth policies that allow us to live, play and work closer to where we live, thereby reducing dependence on long commutes and the need for wide, beautiful courts.

CHAPTER 14

Green Buildings and Urban Planning

Buildings have the ability to reduce a city's urban heat island effect. Urban development that relies on the use of a dark, flat, heat-conducting material such as asphalt to build roads, parking lots, and rooftops absorbs sunlight and holds heat long after the sun has set. To combat urban heat islands, take targeted action in two areas. In addition to making smart choices about the color and material of buildings and paved surfaces, we can also make efforts to increase the reflective properties of the city. For example, buildings can be topped with cool roofing solutions. Cool roofs are an effective and simple means of reflecting the sun's heat. By using high-albedo materials such as metal or reflective asphalt shingles, cool roofs reflect sunlight from the building instead of absorbing the energy. Not only does this reduce the amount of heat that enters the building and is absorbed by the material below, but it also helps reduce the amount of heat that radiates through the air.

As cities and towns expand, harmonizing human habitats with the natural world is a growing concern. The construction industry accounts for nearly 40 percent of global fossil fuel consumption and is one of the highest carbon dioxide emitters within the industrial sector. Buildings also contribute to the urban heat island effect by absorbing sunlight and holding heat long after the sun has set, especially when built using materials such as metal and concrete. Urban planning also plays a role in dictating how resources are used and organized in cities. These decisions affect transportation networks, access to resources, and how we use water and energy. Redesigning green buildings and making green choices in urban planning are necessary steps to protecting the natural world and the limited resources it provides and is a collective response to mitigate the climate crisis.

CHAPTER 15

Climate Justice
and Equity

Our goal is a just and sustainable world, and within that there is no question that the many who are experiencing poverty deserve to move upwards. And future generations, far more late to the planet than any who are currently engaged in helplessness suffer even more. There is much discussion of climate justice and adaptation and of climate equity. As we discuss the allocation of funds, though it is important to remember that the scope of the issue we are addressing is massive. If our understanding of climate justice is to mean anything, it must focus on the poor and what here Pope Francis calls intergenerational justice. That is, all decisions need to appear with an eye as to how they will impact the many who are poor, now, and in the future.

It is to be expected that some are more vulnerable because of where they are located, and in this world, those with no resources are the most vulnerable of all. One measure of an ethical system is how it treats its most vulnerable members. A climate economist has said that if a person living in poverty took one step up the economic ladder, that step would increase that individual's carbon footprint by about 10 times. This is because this person becomes more industrially engaged, accepts heating, and purchases cars. Also, the poor proper because of the activities of those with considerable wealth, living in large homes, flying, and otherwise using large quantities of energy in their lifestyle.

CHAPTER 16

Health Impacts of
Climate Change

Indeed, climate change is often called an environmental health problem. It is, however, by no means the sum total of health concerns associated with anthropogenic interferences with Earth's environmental systems. Climate change is merely a component of a more comprehensive challenge involving a plethora of exposures that contribute to human illnesses. As for the climate impact on the infectious disease spectrum, yes, the changing conditions could cause the spread of tropical and other diseases into new latitudes, yet the migration of those diseases to new territories is not that distinctly different from the spread to new territories that has been manifest for several centuries, promoted by increasing ease of intercontinental travel. Certainly, the last century of climate change has therefore been no driving force in this area, with global urbanization, industrialization, and air travel readily providing sufficient amplification of the risk to obtain such substantial regional alterations. More immediately evident manifestations of the impact of climate change on health come from sentinel ad hoc events.

Weather-related events are generally the first to come to mind, including hurricanes, heat waves, droughts, fires, flooding, and storm surges. The injuries, displacement of people, and interruption of social institutions all promote immediate health concerns. Famously, the first zygote conceived by the Titanic survivors during 1912's April sinking, and born the following February, was named Titanic.

Garry McDougal, MD states, "Just as war is too important to be left to the generals, so too is the health impact of climate change too important to be left to the physicians." However, it is nonetheless useful to outline the direct and indirect health effects of climate change, as they can serve to motivate public action and help generate a passionate response to the call to restore the balance of the natural systems.

CHAPTER 17

Economic Opportunities in a Green Economy

Renewable energy will create 18 million jobs in the next 12 years, according to the International Renewable Energy Agency. Natural solutions are also cost-effective options. According to the World Resources Institute, planting trees and using agricultural best practices to store carbon dioxide would cost only 1/3 of the cost of coal in the United States. More than 100 billion has to be invested in renewables to close the hole to reach the Paris Agreement of 1.5 degrees Celsius of 12 years to prevent irreversible climate change. More than 50 percent of these adaptation funds should be given to projects in small Pacific Island countries and other vulnerable countries which are severely affected by climate change. In comparison, in 2019, more than 71 billion has been invested worldwide to help green energy projects.

In addition, many people think that basic green investments are expensive, but it's actually worth it to do them. If you invest 1 dollar today, you will actually get out about 7 dollars in return. It also reduces the electricity bill for more households and creates more new jobs. There is no excuse for not doing it and there's no lack of funding. We need to stop financing pollution and start financing solutions. For example, if banks, investors and businesses support a 100 percent renewable energy strategy to stop the oil and gas expansion we can keep to the Paris Agreement of 1.5 degrees Celsius in global warming. Renewable helps to provide energy security and creates jobs in the local development. The energy, oil and gas sectors can join to achieve this goal and do something to slow global warming.

According to the World Economic Forum, we have the opportunity to change our economy and create 26-43 trillion and 65 million low-carbon jobs. This clean energy revolution will be the biggest economic opportunity of the century. We have a five-time bigger economy than we have today. This is the kind of revolution we need to not only address the climate crisis but to ensure a fair, resilient, and prosperous world.

CHAPTER 18

Climate Resilience and Disaster Management

Localize the Politics of DRR: To build back better after the pandemic, increased political power at the local level might make it possible to adapt and integrate disaster risk reduction strategies. The current focus of national and international COVID-19 response plans on local stakeholders must be promoted to support the overall recovery effort. This includes the impact of health, environment, trade, and financial support to strengthen coordination and promote local measures. As a result of unprecedented change, unprecedented outcomes, and monetary benefits, funding and development for international cooperation are inadequate. Regional contributions will be indicated by international monetary contributions. Community organization and inclusive governance must integrate women and many vulnerable people to enhance their strengths and capabilities. Localizing decision-making in disaster risk management and risk reduction is important, but empowering women, girls, children, elderly people, and disabled people to be made inclusive and effective in this process while taking into consideration the competing vulnerabilities and needs of different dimensions of diversity, e.g. race, age, and culture.

Development and Climate Resilience: The steps that we take today to respond to the multiple-country COVID-19 crisis are enough to build a sustainable, climate-resilient economy, but they can leave behind lasting consequences. Lower-income countries have limited capacities and resources to manage multi-hazard as well as governance, development, engagement, insurance, and uncertainty. These need to be prioritized as they steer towards a green, sustainable path of recovery from the COVID-19 pandemic. National COVID-19 recovery efforts that include elements of sustainability and resilience must be integrated into national climate plans. In view of the medium to long-term economic impact of the COVID-19 pandemic, investments at the local level can create opportunities to transition to sustainable livelihoods. Equities in risk management should aim to improve overall resilience in the economy (secondary and premature disasters). The aim is to protect and enhance livelihoods and economic opportunities by minimizing losses and maximizing benefits. Investments in disaster risk reduction, risk response, and recovery support growth and development to promote long-term resilience, poverty reduction, and livelihoods that are compatible. It is important to strengthen open spaces for partnerships to promote inclusion and equality.

CHAPTER 19

Indigenous Knowledge and Practices for Sustainability

If we take agro-forestry practices, which are developed in many regions, people cultivate the land with a horizontal system that makes the region green and beautiful. Not only does it make the region green, but it also absorbs four times more carbon than any tropical forests.

Long before, we did not need advanced technology, market mechanisms, or government policies to obtain some products in our daily lives. Production and marketing of the products in society at that time did not produce carbon gases. Why? Because the production of goods was very minimal, so emissions were also very minimal.

In indigenous societies like ours, ecosystems are our homes and our homes are part of ecosystems. That is why we have sustainable environmental management, which has been practiced and developed for generations. This accumulated knowledge and experience has been passed from generation to generation. We have local language and vocabularies for living species in our surroundings, and about 70 of them are in use for managing our agricultural activities.

To address this, we have developed a green school curriculum by jointly designing it with 3,000 traditional architects. Now, about 6,000 teachers are using the curriculum located in 6,000 schools across the country.

In Indonesia, the green economy approach is a part of our culture that our ancestors have been using and is embedded in our traditional ecological knowledge and its practices. Our traditional houses are built with the concept of harmony between humans and the earth and environment where the house is located. We believe it should be preserved as part of the national charm.

In the face of global warming, which is touching every aspect of our lives, most nations have adopted the policy of addressing this crisis by embracing a green economy, using renewable sources such as the sun, water, and wind. We have started realizing that our buildings must now be designed with a natural source of energy and hence must respond to our local climate and culture.

CHAPTER 20

International Cooperation and Collaboration

Scientists who compared climate impacts of bioenergy with carbon capture and storage and of solar radiation management both viewed the proposed risk reduction strategies with caution. An estimated geo-engineering cost of $100 to $400 billion per year carried implications that global solutions would be market-led and would preempt evolve in small, affordable increments. By contrast, the cost of bio-energy with carbon capture and storage was, they suggested, entirely predictable in that it could have no impact but could consume $1,800 billion in development funds over the next two decades. In earlier studies, Pea, L., and the UK Scientific Alliance concluded that in Dalby and Morita, H., High hopes for accelerated carbon reduction in the EU seemed illusory, with industrial relocation to non-restrictive partners masking the short-term European carbon benefits to only a 1 percent reduction.

International cooperation is also crucial to the climate crisis, however difficult to achieve within a political and economic system that encourages competition, military posturing, exploitation, and disregard for laws, much less views of the earth as the wellsprings of life. Factors now contributing to climate change are profiteering and heedless extractivism from topsoil to stratosphere. Disease, diminished resources, damage, and impoverishment run unchecked; lawlessness abounds and underpaid guards the frontiers. Capitalism, many have noted, underwrites such behavior when production and labor for profit alone are the guiding metrics, each of these without regard to consequences. World-wide there's a need for massive switch of financial resources to meaningful purposes, yet the climate change that demands it also diminishes the wealth that could "fund" such re-allocation.

CHAPTER 21

Climate Change and Water Resources

The climate crisis provides a fundamental signal to manage the water: when future inflows are expected to be altered and thus water management is essential, then culture, fickle governments, and flesh-kneading hopeful individuals can articulate long-term agreements and build a unified strategy. Countries, states, reservations, cities, political parties—even the famously individualist ranchers and farmers of the West—can pull away from discord and toward a more harmonious use of detritivorous resources. Even capitalists recognize the politics of self-interest. Often, it is our birthright to tap into the competitive Darwinian engine. Each law of self-interest is untwisted.

If the millennial United States—the modern arid West with its wildfires, its vast irrigated agriculture, its cities, and its wildlife—were to be seen as a body, the water crisis would be the liver: over-sized, inefficiently used, and deeply affected by what is imbibed. According to parched officials of the West—politicians, farmers, entrepreneurs, and tribal leaders—they must manage their scarcer water as a body does its increasingly fatty, problematic liver. What incentive would be so powerful that the nation, that most impressively self-indulgent and competitive of bodies, would collectively manage the West's problem liver, this most essential and ill-used resource, toward useful health? The climate crisis delivers that incentive.

CHAPTER 22

Sustainable Consumption and Waste Management

Would a global reduction in material consumption lead to reduced CO_2 emissions? The answer, according to scientific calculations and models developed, has shown that this path is indispensable and unlikely. This does not necessarily profess the absolute failure of production and consumption, but a new paradigm of it. Waste management is one of the major problems that must be solved at the community, national, and global level. Different levels of society - governmental and non-governmental institutions, industry, agriculture, commercial and public service companies, as well as individuals, must be involved in the process of developing global strategies to solve this problem. Experts involved in waste management will ultimately require a more responsible burden on the producers, together with a more rational attitude towards the use of materials and related waste generation. It is crucial to educate citizens about the principles of waste generation, consumer culture, and how to deal with waste management in order to minimize its environmental impact. Highly sensitive to the consumer revolution, it is sufficient for the population to realize that their daily habits are potentially leading to the depletion of natural resources and causing environmental pollution.

In the last few decades, the widespread transformation of production methods and the consumerist culture has taken consumption to a central position in the generation of environmental problems. The increase in consumption of goods and services leads to an increase in the emission of greenhouse gases. Thus, adopting sustainable consumption patterns can produce lower carbon emissions. Actions to inhibit consumerism can be seen as inappropriate and an infringement on the rights and freedoms of an individual. Precisely for that reason, governments need to make adjustments and adopt public policies that will achieve incentives. The first path to adopt is to get in with the most apparent levers to impulse a change in consumer habits - high taxes on inappropriate products and the reduction of them on other sustainable solutions.

CHAPTER 23

Climate Change Communication and Media

The media needs to include students, everyone working in the media sector, as well as influential and supported lay people in disseminating the problem of climate change. First, they should be made to understand the huge burdens that all citizens will face from the increasing variation, which is expected to become more unpredictable and difficult to manage as the current political failure continues. They must realize that we must act now or face an unprecedented social, economic and environmental crisis in our history. Various methods of increasing public awareness and advocating for greater involvement can be tried. If the public is not clearly aware of the two alternatives, it will continue to form opinions and take actions that will make it impossible for us to live in a prospering society. Educating the public about real-world solutions, proactive measures, and information on a global scale that provide alternative data to increase the number of members in local and social communities must be pursued.

The nature of climate change makes it difficult to communicate. People don't realize that it's not a long-term problem, it is already affecting us. Furthermore, people tend to feel that changes in weather are local and restricted to certain seasons of the year. They do not feel like something is changing unless the temperature gets extremely high or extremely low. The recent increase in contrasts implies harsh winters and cruel summers, making some people feel, if they are verbotten, the opposite that is the sign that the climate has not continued to stabilize and that everything is a story of scaring the public (manufactured crisis).

CHAPTER 24

Climate Science and Research

Global studies (which are growing in importance) include non-theoretical preparation, studying geographic statistics, where the main problem is poor and lack of reliable databases applicable to long-term data. But a number of local studies can be expected to control this region by the new pre-existing geographic conditions. The concerns arose that the observations did not have long-term thermodynamic calibration from globe to the point at which the data could be selected. The outcomes of the research led to the conclusion that 4 years seemed unlikely to achieve such. Some effort would be made to improve the records used in the models and applications of the global databases forth. The records pertaining to the last 30 years showed a close agreement between the observations, extended over the same period, of a comprehensive sample of the crops used. They were made to identify outliers and possible differences between the observed data that could be accounted for in implementing methods to adjust records, such as those due to the time of observation and instrumental changes.

Recently, I read an article in which a scientist confirms the necessity to keep researching everything about climate change to better understand the problem and to find necessary solutions. Climate science is a growing field of scientific research related to the climate. It refers to the concentrated efforts to monitor and understand the structure and behavior of the climate, especially regarding weather patterns, short and long-term changes. Climate science is an interdisciplinary research, drawing on various fields including climatology, chemistry, biomechanics, physics, electronics, oceanography, geology, geothermal, and mathematics. It is mainly concerned with the research area.

CHAPTER 25

Climate Change and Human Rights

There can be no winners on the planet, or among its citizens, when the climate crisis undermines all rights. Climate change is driving sea-level rise through two main processes: the thermal expansion of seawater as it warms and the melting of glaciers and ice sheets. This colossal melting of glaciers and the ice sheets that exist at the Poles is causing 30 sea levels to rise, threatening coastal and island-dwelling 31 communities and contributing to the migration of people due to environmental reasons. The consequences of climate change can thus amount to a vast range of negative impacts on individuals, including legal, physical, economic, or political factors that put a spotlight on the importance of enhancing climate resilience. Clearly, this would help the effective protection of human rights, as well as implementing actions for sustainable development.

The climate crisis will impact human rights in almost every aspect. It is very important that we show how global heating will affect the population and what needs to be done to avoid these worst-case scenarios. Erosion of people's safety, their right to food due to disaster-induced damage of agricultural production infra-structures, their right to health due to extreme weather events, and loss of access to drinking water. There was water, but the 15 bar - the equivalent of a heavy rain shower - caused our entire mess to collapse and the entire house started to slide down the mountain - inside the home, too. I felt that on my feet. Many landslides occur when intense rainstorms drench soil that has already been stripped of its stabilizing vegetation by cutting down trees or when earthen slopes are weakened by fracking.

CHAPTER 26

Climate Change and Gender Equality

Exclusion, including socio-economic disadvantage and political marginalisation, remains a factor which increases a woman's vulnerability in a changing climate. The geographical location of the woman and/or her ethnicity often exacerbates the full effect of the economic poverty she is experiencing. However, in addressing women's roles as "vested ecological protectors" with limited economic choices, solutions must also include the concept of the gender division of labour. Women, who have few to no economic opportunities to engage in decent work or earn money, are often primary caregivers, food providers, and water collectors. Additionally, they do family recycling and processing of organic materials, including animal dung, often in dangerous, climate-intensive environments. Women have a right to protection and to work within conditions which preserve their health, dignity, and integrity. It has been sought from the Local Agenda to the Bali Action Plan to the Paris Agreement. However, the official climate change policy space has yet to analyze the gendered division of labour, as per the Bali Action Plan, and is still lacking the impact of 'actions to advance gender equality', as found in the Paris Agreement.

2018 saw an increase in the gender dialogue versus climate change. The United Nations COPAS held an "activist" Gender day, and an "eco-feminist" themed one-day event, as did WEA Paris. Women's perspectives and rights and women's vulnerabilities were, once again, showcased. Recognition and a declaration of a "gender and climate justice moment", however, are not enough. Effects, measures, and energy policy decisions are not only sidelined but are actually perpetuating women's rights' violations and violence. These include cases where villagers face threats as well as sexual and physical violence from companies building things like coal-fired power plants, oil and gas pipelines and other energy infrastructure, such as dams.

CHAPTER 27

Youth Engagement in Climate Action

Question: I have talked to my friends to make them aware of climate change and helped plan climate strikes at my school. What else can young people do to get adults to act on climate change when they have not listened to scientists and conservationists?

Answer: In my book, I tell my personal story about how I felt despair about the climate crisis but found hope by getting engaged in action. I also describe what I've learned about climate science and solutions and provide specific suggestions for actions that individuals and families – including young people – can take to help address the climate crisis. It was very important to me to write in a way that would be accessible to young readers, so I focused the writing for clarity, and I provide explanations for terms that are unfamiliar to most 7th-graders.

Question: You recently published a book, "What Can I Do? My Path From Climate Despair to Action." What do you hope the teenagers who read your book take away from it?

In November 2021, a live broadcast with hundreds of teenagers took place to talk about how young people are leading the way in the fight against climate change. Here are their questions and my answers. Thanks to all for the thoughtful conversation.

CHAPTER 28

Philanthropy and Funding for Climate Solutions

Ongoing support of effective climate policies and the funding of innovative renewable energy production and energy efficiency projects can further increase the rate and optimize the transition to a clean energy future. Donors are important agents in the process of engaging scientists and industry to develop the future clean options that are as yet unrealized. Foundations can also help the process of transforming laboratory and pilot plants as they develop discoveries into companies capable of leading and financing the construction of the commercial scale projects. The more we can have any new technology ready for a time when it can provide a significant carbon reduction, the more effective the timing will be.

Philanthropy has a critical role to play in addressing the climate crisis. By providing funding, engagement, education, and knowledge to fuel climate action, philanthropy can make a significant impact on the causes and consequences of climate change. Recognizing the urgency of the problem and the opportunities to address it, many faith traditions are also engaging with philanthropic partners in deepening their approach to caring for creation. Through careful and informed investment in a flourishing planet, we can inspire governments, corporations, and citizens the world over to meet the challenge that is before us.

CHAPTER 29

The Role of Businesses in Addressing Climate Change

Patrons of the planet as third-party companies of planetary restoration: experts advise leveraging current state-of-the-art technology for verified and quantified collective human activity and restoration efforts as the best and most achievable methods of improving climate damage all over the world. Seventy-five percent of the major economies already have net-zero commitments, and almost 60% of fossil energy must be phased out by 2030 to 80% by 2030. In addition to capturing carbon and restoring the earth beyond guidance thresholds, these projects create more valuable results such as human employment, habitat restoration, and corridors for flora and fauna that are essential to overall well-being. Finally, with the beneficial impact provided by digital action platforms and local community responsibilities, all of us, who cannot accompany our hikes from countless extinguishers into real earth sanctuary remains, to listen to a spectrum of moors, provided we build a profound climate present and future. We completely on was the planet, making sure the sharing platforms of the EPA or the cleaning system of the environmental sewage compartment can help regenerate our eco-dollar-resiliency system currently underway.

In 2019, we used more resources than the Earth can produce - running up about $4 trillion in ecological debt. The cost of dependency on unsustainable food, fuel, and fiber is not just a climate one - a first-ever accountant's report shows that inaction in the natural world could burden the global economy at $10 trillion per year. That's why for the first time, experts are saying we can no longer address climate, biodiversity, and pollution crises in silos - we have to address them together. What's more, less than 1 percent of 8,000 companies worldwide report comprehensive progress on the majority of nine key human rights, environmental, and anti-corruption issues. Meanwhile, every single industry is contributing to nature loss beyond planetary boundaries while failing to meet nine out of the ten United Nations Sustainable Development Goals considered. In this chapter's second in-depth look at climate solutions in 2021, we explore nature-positive industries and carbon-positive businesses of the present and future. Collective forest-boundary restoration has been noted to "provide the most cost-effective opportunity to combine the most achievable levels of carbon removal with strong solstice benefits."

CHAPTER 30

Ethical and Moral Dimensions of the Climate Crisis

Based on such reasoning, it becomes important for everyone to start thinking about how ethical and moral considerations relate to their decisions and actions regarding what to eat. Climate change will not be solved by various trick solutions to increase food production or high technology or industrial strategies to reduce greenhouse gases. Until we learn to respect life, Earth will not heal. If the hazards of climate change as well as the destruction of our agricultural environment are to be avoided, reliance solely on pressed high technology solutions and expanding features of non-sustainable agriculture practices is as morally unjustified and unjust as exploitation of citizens. There is something in human nature, deep in our way of thinking and feeling, that craves something better than an endless consumer lifestyle. We want to live satisfying lives of fulfillment.

Climate change involves many types of ethical and moral issues. This civilization-wide human predicament implies the urgent need for new ethical and moral paradigms or understandings of our places within society and the world. Endless consumption of material items alone fails to fulfill genuine human needs. Many come to realize the profound and diverse ways human actions contribute to environmental disruption. Consumer demand for meat belies the hidden dark side of its agricultural origins regarding the treatment of food animals and human society. Given these circumstances, what are the moral implications? We suggest that these issues involve a consideration of the unity and moral implications of life that is too frequently ignored in modern society.

CHAPTER 31

Hope and Resilience in the Face of Climate Change

We create hope and resilience when working together to protect what we care about. One way to protect our hearts in this era of climate disruption is to protect the parts of the world we love, and that inspire us – its creatures, lands, waters. This impulse is often most fruitfully explored locally. Sometimes, to protect what we have, as experienced during his protest out in Utah, we need to protect what we don't yet have – in his case, the possibility of future generations ever enjoying a similar view of the mountains. The danger to the Colorado plateau encouraged many to try to protect that land; doing so is a way to find resilience and hope, a practice to revitalize ourselves and our communities.

Whether or not you've been personally affected by extreme weather events, it can be painful to witness climate chaos. These trends and disasters can trigger feelings of loss, trauma, and despair. But, recognizing that same ecological damage may also be the incentive to cultivate personal and collective resolve and resilience in the face of the crisis. What are the elements of a coherent ecology of hope? As scientists, spiritual guides, and wisdom traditions have all noted in recent years: we need not just a shift of policy to swiftly transforming the fossil-fuel economy into one based on sustainability and social justice, but also a profound shift in our hearts that brings us wisdom and resilience in the face of the crises of climate and justice.

CHAPTER 32

Conclusion

As I inhabit new identities and lead unconventional lives, as I face forward into my unpredictable future, as I strive towards embodying the illumination of the sun, the sanctity of intelligence, the splendor of strength, and the passions of love, I appear to you transformed. Galvanized. Joyful. Hopeful. But my mission isn't about me. Yes, in the journey of my own self-discovery and empowerment, I have become the woman I needed to be in order to live healthily, speak truthfully, love unconditionally, and live sustainably in a chaotic and distressing world. However, in my journey to define success, hope, and failure to further include all of humanity but the Earth, I am just getting started. I am now ready to hold creative, powerful, and heroic space for others—profoundly and intentionally.

Throughout my life, I have clung to the illusion of control. By sacrificing my limitations, weaknesses, and fears at the altar of conformity, comparison, and self-image, I mistakenly sought to exert control over my life and well-being. Following this rigid blueprint seemed like the only way to survive, succeed, and find happiness. Yet after living these definitive years, I recognized the glaring problem of this costume: It was based on an artificial understanding of power, intelligence, beauty, and success. This unnatural facade was designed to protect me from the hard work of exposing my own humanity. By dedicating countless energy and suffering towards achieving societal ideals unattainable by either myself or others, I placed my identity, worth, and sense of peace at jeopardy. Coming to terms with my cluttered, brittle, and eccentric self, awakening a DHP, I have discovered untapped reservoirs of hope, love, and sustainability.

Milton Keynes UK
Ingram Content Group UK Ltd.
UKHW031400011224
451790UK00009B/133

9 798330 582556